I0062283

Bayes Theorem

The Ultimate Beginner's Guide to Bayes' Theorem

by Arthur Taff

TEACHING NERDS
TN

Copyright © 2017 By Arthur Taff
All rights reserved. No part of this book may be reproduced in any
form without permission in writing from the author. No part of this
publication may be reproduced or transmitted in any form or by any
means, mechanic, electronic, photocopying, recording, by any
storage or retrieval system, or transmitted by email without the
permission in writing from the author and publisher.
For information regarding permissions write to author at
Arthur@teachingnerds.com.
Reviewers may quote brief passages in review.

Arthur Taff
TeachingNerds.com

Table of Contents

Who Is This Book For?

Bayes theorem describes the probability of an event based on other information that might be relevant.

Essentially, you are estimating a probability, but then updating that estimate based on other things that you know.

This is something that you already do every day in real life.

For instance, if your friend is supposed to pick you up to go out to dinner, you might have a mental estimate of if she will be on time, be 15 minutes late, or be a half hour late. That would be your starting probability.

If you then look outside and see that there are 8 inches of new snow on the ground, you would update your probabilities to account for the new data.

Bayes theorem is a formal way of doing that.

This book is designed to give you an intuitive understanding of how to use Bayes Theorem. It starts with the definition of what Bayes Theorem is, but the focus of the book is on providing examples that you can follow and duplicate.

Most of the examples are calculated in Excel, which is useful for updating probability if you have dozens or hundreds of data points to roll in.

What Will This Book Teach You?

Bayes' Theorem contains an essential formula we use to determine conditional probability.

With a thorough understanding of the theorem, one can not only compute a number of real life probabilities, but also get a glimpse of how we use probability and statistics, an essential mathematical topic in the modern world.

This book provides definitions, explanations and easy to follow examples for individuals of various mathematical levels. After going through this book, the reader learns what Bayes' Theorem is, when it is applicable, and how to apply it.

Introduction

Probability is a concept that appears everywhere in our lives. It can give just about anyone a little more insight on how events occur every day, whether they are random or not.

Say you want to take a walk, but the weather man says there is a 60% chance of rain. Maybe this particular weather man is only right 70% of the time. Do you still want to take a walk if it means possibly getting rained on? Maybe you don't care, or maybe you don't want to get your new shoes wet. Maybe you think it's a hassle to bring an umbrella.

Aside from adding convenience and insight to our everyday lives, understanding probability also allows us to excel in numerous fields of study, including forecasting, medicine, finance, and law. You may want to know whether taking a walk is worth getting rained on, but whoever actually predicts the weather needs to thoroughly know the statistics, or else he'll be out of a job.

Weather prediction is just one example of the infinite applications of probability theory. Let's say I've been feeling sick for a couple days. I have a job interview on Thursday. Today is Monday, and I want to make sure I'm healthy by Thursday, but I'm not sure if I can afford the time and cost of going to the doctor before then. Based on the history of my illness patterns, I have a 20% chance of still being sick on Thursday. Pretty decent odds of being healthy, right?

However, one of my coworkers might be sick, and this can affect my recovery negatively. Or maybe I'm on a new health kick that can positively affect my recovery. Maybe this new health kick hasn't impacted my immune system yet. Events related to my recovery can affect the probability that I am still sick on Thursday, and these events should influence my decision to go to the doctor before then. Maybe I learn that going to the doctor is only 40% effective for the illness I have.

Again, there are a number of factors that can influence my decision, and we can account for these factors by calculating conditional probabilities.

This may seem like a whole mess of numbers and different probabilities. However, with the right understanding and practice, we can apply Bayes' theorem to solve problems for true conditional probabilities.

For simple examples, the probabilities can seem fairly intuitive, but using Bayes' equation gives us an exact numerical value for conditional probabilities. Furthermore, we can use our applications of the theorem to confirm our intuitive thoughts surrounding conditional probability.

The Formula and Notation

Here's the formula we'll use for basically the entire book.

$$P(A \mid B) = \frac{P(B \mid A)P(A)}{P(B)}$$

Where

$P(A|B)$ is the probability of event A given that event B is true,

$P(B|A)$ is the probability of event B given that event A is true,

$P(A)$ is the probability of event A without regard to event B,

$P(B)$ is the probability of event B without regard to event A.

P(A|B) and P(B|A) are referred to as conditional probabilities because they calculate the probabilities of A and B, respectively, based on conditions that may be related. Remember, Bayes' theorem does not apply to events that are unrelated.

For example, if Johnny and Sally are walking to work on separate routes and Sally forgets her purse, Johnny will not be late to work, because what Sally does is not related to Johnny's walk to work.

To use the formula to solve for P(A|B), all we need to do is determine what A and B are, then figure out P(A), P(B), and

P(B) assuming A occurs, or P(B|A). Once we have these values, the substitution is simple and we can successfully apply Bayes' Theorem to find the probability of A given B.

Bayes' Theorem Explained: 4 Ways

Here are a few ways Bayes' Theorem can be explained:

1. Bayes' Theorem helps us update a belief based on new evidence by creating a *new belief.*

2. Bayes' Theorem helps us revise a probability when given new evidence.

3. Bayes' Theorem helps us change our beliefs about a probability based on new evidence.

4. Bayes' Theorem helps us update a hypothesis based on new evidence.

The only problem? Applying the theorem is not intuitive, at least not for most people. This is where visualizing a problem that entails using Bayes' Theorem can be a BIG HELP.

Conditional Probability

This concept is commonly introduced in elementary statistics. The occurrence of some events is sometimes dependent on others. This is to mean the likelihood of one event occurring may be more or less depending on how other events occur.

For example, the probability of a football team winning a match is dependent on that particular team scoring goals. Another example is the likelihood of causing a fatal accident is dependent on your driving speed etc. we cannot exhaust the number of life examples.

Now let us get the formal definition of conditional probability.

Remember,
$$P(A) = \frac{N_A}{N}$$

Let us now consider one additional event B, we will concentrate on the condition that event A occurs only when event B occurs. To be able to define this probability we will perform a set of trials and record the number of times B occurs.

We have, $\frac{N_{A \cap B}}{N_B}$

Where N is the number of trials, the probability of these dependent events occurring is,
$$P(A \cap B) = P(B) \cdot P(A|B)$$

If we divide through by P (A) and commute the whole equation, we get
$$P(A|B) = \frac{P(A \cap B)}{P(A)}$$

This is the conditional probability formula, and when read it sounds, the probability of event A given event B. this will be of great help as we derive Bayes theorem

Example3.1

A class two teacher gave his pupils to questions, 42% of the class did well in both questions and 25% performed well in the first test. What is the percentage of pupils who performed well in the first also did well in the second.

Solution

Let A be passing the first test and B be passing the second test, therefore

$$P(B|A) = \frac{P(A \cap B)}{P(A)}$$

$$P(B|A) = \frac{0.25}{0.42}$$

$$P(B|A) = 0.60$$

And remember we are asked to give our answer as a percentage, therefore

$$P(B|A) = 0.60 (100)$$

$$P(B|A) = 60\%$$

Read This FIRST - 100% FREE BONUS

FOR A LIMITED TIME ONLY – Get the best-selling book *"5 Steps to Learn Absolutely Anything in as Little As 3 Days!"* by Edward Mize absolutely FREE!

Readers who have read this bonus book as well have seen huge increases in their abilities to learn new things and apply it to their lives – so it is *highly recommended* to get this bonus book.

Once again, as a big thank-you for downloading this book, I'd like to offer it to you *100% FREE for a LIMITED TIME ONLY!*

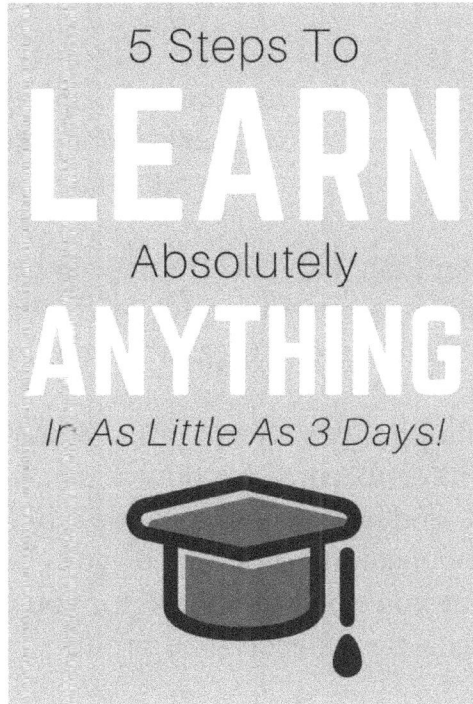

To download your FREE copy, go to:

TeachingNerds.com/Bonus

What A Test Is and Isn't

"Grown-ups love figures," lamented the eponymous hero of a beloved literary work. That seems to be true years before and years after the said literary work was published.

You must be familiar with such reality. When you make a decision, you always consider a lot of factors. More often than not, those factors are expressed in numbers.

Take grocery shopping as an example. When you buy food products, you have to consider how much you need, how much they cost, and how long the goods will last. If you are on the fitness-conscious side or if you have a dietary requirement that you need to adhere to, you also weigh on the number of calories in each food product.

The same thing goes with shopping for an appliance. Aside from the price and lifespan, you consider the size, weight, and level of energy efficiency. If you shop online, you rely on product specifications and consumer ratings.

When you watch or read the news, you are flooded with large numbers and percentages as well. Millions, billions, 90%, 80%, 3%, 1% and so on are used to denote wealth, severity, efficiency, satisfaction, economic growth and other aspects that must be quantified to get a better grasp of the news.

Some people just take every bit of information handed to them. They take everything as the truth and so they make such information their new beliefs. They hold on to such beliefs and make them the bases for their decisions.

What if they misinterpreted the information? Worse, what if the information they took was based on an erroneous test?

Skepticism Goes a Long Way

It is a good thing to be skeptic every now and then. Whether you are a little or largely skeptic, you can spare yourself from some of the disadvantages of misinformation.

If you are skeptic though, the optimists will often tag you as too doubtful or too pessimistic. You could hardly persuade them to believe your claims about the unreliability of some information being spread around.

Fortunately, there are still realistic and open-minded people out there. They are the ones worth encouraging into questioning dubious information that are heavily publicized.

Or, you may just be a skeptic for your own sake. If you are going to question information, do not just focus on what is presented and how it was intended to be used. You should also weigh on how the information was collected. The process is often a test and the results of such test serve as the bases for information.

Why Test

Tests are done in the development of products, services, equipment, treatment, marketing strategies, standard operating procedures, and solutions to a plethora of problems. Tests are intended to predict outcome. In statistics, such outcome is referred to as probability. Probabilities are often expressed in percentages.

Knowledge on probabilities is vital in the preparation for the worst-case scenarios. Medical tests are a great example for this. By going through medical tests, doctors may be able to pinpoint your health condition and provide proper treatment.

Aside from preparation, probabilities also help professionals such as manufacturers, designers and strategists in the improvement of their products, services, designs or plans. Take client surveys as an example. Service providers use survey as a test to find out whether their work is satisfactory or unsatisfactory. If the survey suggests that their work is unsatisfactory, the service providers have no other choice but to improve.

However, a lot of tests that people—professionals or not—rely on are not that perfect. No matter how noble the intent, how meticulous the implementation, or how proficient the tester, tests are still flawed. Even some of the standard scientific tests and measuring equipment have innate rates of errors.

The Fault in the Tests

Tests can be inaccurate. They can detect things that do not actually exist or they cannot detect the things they are supposed to spot.

In medical statistics, there are concepts that refer to such inaccurate results. These concepts are false positive and false negative. A false positive is a result in a medical test wherein the patient tested positive for a disease even though he does not actually have it. In contrast, a false negative result entails that the patient tested negative for a disease even though he actually has it.

In contrast, accurate test results are referred to as either true positive or true negative. As the name suggests, a true positive result means that the patient tested positive for a disease that he has while a true negative result means that the patient tested negative for a disease that he actually does not have. True positive, true negative, false positive and false negative results are known collectively as the four types of probabilities.

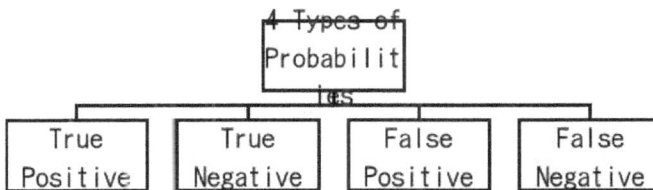

```
            ┌──────────────┐
            │ 4 Types of   │
            │ Probabilit   │
            │     ies      │
            └──────┬───────┘
   ┌───────────┬───┴───┬───────────┐
┌──────┐  ┌──────┐  ┌──────┐  ┌──────┐
│ True │  │ True │  │False │  │False │
│Positive│ │Negative││Positive││Negative│
└──────┘  └──────┘  └──────┘  └──────┘
```

What's in a Name?

Each test has its own name. Some are named after the one who devised the test. (Examples include Turing test and Rorschach test.) Some are labeled according to their process. (Car emission test is an example.) There are also tests named after the object or condition they are supposed to detect. (Examples include pregnancy test and drug test.)

The names of tests do not actually matter. The labels are just given for convenience. They do not have any bearing whatsoever on the outcome nor establish the accuracy of the test. You can even give nicknames to all the tests you know.

The tests are not about the names anyway. The true meaning of a test is based on its accuracy and flaws.

Test versus Event

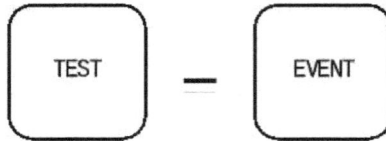

To the non-discerning eyes, the test and the actual event are the same thing. The actual event refers to the object or condition that the test is supposed to detect. For example, in a pregnancy test, the actual event is pregnancy. In the case of alcohol test, being drunk is the actual event.

Unfortunately, the test and the actual event are distinct from each other. Therefore, the proper diagram for the relationship of the test and the actual event should be the one below. The probabilities that tests can offer are not the real probabilities for the actual event. Later on, you will find out how the test probabilities and real probabilities differ from each other.

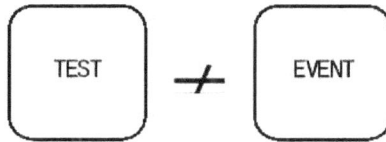

Tweaks to the Tests

Inaccurate test results can be dangerous, costly and worrisome. It is only fitting to know the flaws, understand their severity and correct them before the test is considered and implemented with utmost confidence.

One of the steps towards correcting the errors in tests is weighing all the probabilities.

A common cause of misinformation that even some professionals are not aware of is the use of true positive results as the sole basis for the accuracy of a test. The true negative, false positive and false negative results should be considered as well.

People like to read natural numbers more than percentages. Therefore, professionals should properly and carefully interpret and convert probabilities into actual numbers.

The Bayes' Theorem helps you in doing so. If there is Pythagorean Theorem in geometry, there is Bayes' Theorem in statistics. What the later theorem does is that it enables the user to find out the more accurate probabilities for an event.

If you are among the grown-ups who love figures, Bayes' Theorem will be one of your best friends. The theorem also assists you in taking your skepticism to a whole new level.

Simple Bayes' Theorem Application

Imagine that there are four cups of the same shape and size in front of you. There are two blue cups and two red cups. The two blue cups plus one red cup contain orange juice while the remaining one red cup has water. Then, you are blindfolded and asked to pick a cup and drink it. You choose one and the content tasted like orange juice. What do you think is the color of your chosen cup?

You are likely to pick blue cup, right? That is a logical choice considering that there is a 2/3 or around 66.67% chance that the orange juice is from a blue cup. That is greater than the 1/3 or 33.33% chance that the orange juice is from a red cup.

You assess probabilities like that on a daily basis. But did you know that you are already applying Bayes' Theorem? The assessment of chances in the problem above is actually a simple application of the theorem.

The Story Behind Bayes' Theorem

Bayes' Theorem also comes by the names Bayes' Rule, Bayes' Law or Bayesian Reasoning. These names all connote to the theorem that tackles the two conditional probabilities of an actual event. The man largely attributed to the theorem is Reverend Thomas Bayes.

A Little Background About Bayes

Thomas Bayes was born around 1701 in England. His father was Joshua Bayes, a London Presbyterian minister. The Bayes family was known to be affluent and they belonged to the sect Nonconformists. Nonconformists did not have their actual birthdays recorded to avoid discrimination.

Thomas studied theology and logic at the University of Edinburg in 1719. After three years in college, he went back and helped his father in managing the latter's chapel.

Around 1734, he moved to Tunbridge Wells, Kent and became the minister of Mount Sion chapel located therein. He earned the title reverend for servicing in the said chapel.

He lived and preached in Kent until he died in April 7, 1761; he was 59 years old. His remains were buried in Bunhill Fields cemetery in Moorgate, London. The said burial ground was also the place where other Nonconformists were laid to rest.

Bayes' Works

Before his death, Bayes only had two published works. The first one was a theological study published in 1731. His second work was originally published in 1736. It was intended to defend Isaac Newton's work on calculus against George Berkeley's critique.

He had no other published works after that so many assumed that his second published work was the reason he was elected as Fellow of the Royal Society of London in 1742. The said group was the premier group of scientists at that time. Surprisingly, it still exists up to this day.

What about the basis for Bayes' Theorem? His work on probability theory was never published. Upon his death, the manuscript for his probability theory was given to his friend, Richard Price.

Bayes-Price Theorem

Like a true friend, Richard Price looked at Bayes' unpublished work and edited it. Aside from rendering his proofreading skills, he also added an introduction that explained the philosophy behind Bayes' theory on probability.

Price's efforts did not go to waste initially. In 1763, the edited version of Bayes' work entitled, "An essay towards solving a Problem in the Doctrine of Chances" was read at the Royal Society. This earned Price the election as Fellow of the said group.

There was a claim that Bayes was not the first one who discovered the theorem on probability. However, such claim was countered a long time ago. Both Bayes and Price are deemed as the twosome behind the theorem. Even though he was not with Bayes when he first worked on the theorem, Price made a lot of significant contributions to it. It is only fitting to name the theorem as Bayes-Price Theorem.

But then again, the name Bayes' Theorem got stuck. Price's introduction served as the basis for Bayesian Statistics but he did not receive much credit for it. Attempts to change Bayes' Theorem and Bayesian Statistics to the more suitable names were futile.

Bayesian Interpretation

The work on the theorem does not stop with Price. In 1774, Pierre-Simon Laplace, a French mathematician, further developed the theorem although he was not fully aware of the existence of Bayes' initial work. Laplace provided the interpretation which was later on called Bayesian Interpretation. This was the interpretation being used up to this day.

Price and Laplace deserve due recognition for their contribution. Nevertheless, you can do that on your own terms. Perhaps after completing this book, understanding the basics of Bayes' theorem and becoming a Bayesian yourself, you will also feel grateful to Bayes, Price, Laplace and others who were not properly credited for their contribution to the theorem.

What the Theorem Is All About

The main objective of Bayes' Theorem is to convert test probabilities into real probabilities. The theorem offers a holistic approach in weighing probabilities; that means it considers true positive, true negative, false positive and false negative results in a test. With such approach, you can find out and interpret the real probabilities for an event.

If the real probabilities are way too low, then it is likely that the test has a lot to improve on. The theorem helps you uncover the flaws in a test. Once you find out the flaws, you will have an idea on the likelihood of false positive and false negative results. Instead of panicking, you can prepare for the disadvantages from such inaccuracies. You have the option to book for additional and more accurate tests to further establish probability.

The use of Bayes' Theorem prevents you from making hasty generalizations as well. You do not just depend on the statistics given. You will try to relate all probabilities. You will try to ask the right questions to get the most sensible answers.

For instance, someone says only 5% of happy people are rich so that means wealth cannot bring happiness. A Bayesian asks: What is the percentage of rich people who are happy? If that 5% of happy people who are rich are 90% of all rich people, then you can argue that wealth may bring happiness.

The theorem can also be used in almost every field: medicine, production, real estate, economics, etc. As long as statistics and probability theory are applicable, then the theorem is certainly applicable.

Bayes' Theorem helps you process and accept new beliefs faster as well. Instead of just taking every bit of information offered, you question the sources and you quantify your doubts. Once you computed the probabilities yourself, you will not end up believing and holding on to false information for a long time.

Beliefs

Your beliefs are composed of all the information you accepted as the truth. You need to update such information accordingly to make wise decisions. But how do you accept new beliefs and update the old ones?

This is when tests become handy. The results become evidence that you use to accept and update beliefs. As to the flaws of the tests, you can use the Bayes' Theorem to assess the reliability of the evidence. The question now is: How do you use the theorem?

Anatomy of A Test

Before learning the Bayes' Formula, you need to know the anatomy of a test first. Each test requires three pieces of information. These are: the prior probability and the two conditional probabilities.

Prior probabilities are collectively called priors. A prior probability is the original fraction that covers all the individuals or objects in a test. In the rich and happy people example above, the original fraction is 5/100.

A minimum of three studies is one of the requirements for the establishment of a prior probability. After the publication of the studies, the American Association for the Advancement of Sciences (AAAS) will then decide whether to accept or reject such prior probability. For the other priors, you can refer to the Handbook of Chemistry and Physics.

The two conditional probabilities are the flaws of the test. The result of the Bayes' Theorem is called revised or posterior probability. It is also the normalized weighted average. You are going to use the posterior probability in updating your old beliefs or accepting new ones.

Terminology

In some examples of the application of Bayes' Theorem, you will encounter terms such as probability and chance. These two terms mean the same thing. Some may also refer to the actual event as hypothesis and the two conditional probabilities as evidence.

Misconception

You do not need to be a professional statistician to understand and apply Bayes Theorem. It is not just for professional or business-related tasks. It is also applicable for personal decision-making tasks.

Exercise 3

- A box contains white and black small balls. Two balls are randomly selected without replacement. The likelihood of choosing a black ball first then white is 0.34, the likelihood of taking a black ball on your first draw equals 0.47. What is the percentage of taking a white ball on your second draw if the first ball draw was black? 72

- The probability that a student is absent and it's on Friday is 0.03, what is the probability its Friday given the student is absent? 0.15

Bayes Theorem for Beginners

Reverend Thomas Bayes, originally stated this theorem and hence its name. You will find most books calling is Bayes theorem or Bayes law. This theorem can be defined as a rule to update prior/initial probability P (A), into a posterior/revised probability P (A|B), taking into account the current updated knowledge available.

In the previous topic, we denoted conditional probability as P (B|A), which is the likelihood of B occurring given that A occurred. The formula for conditional probability was given as

$$P(B|A) = \frac{P(A \cap B)}{P(A)}$$

In this chapter, we will extended our knowledge on conditional probability, so as to get a better understanding of Bayes theorem. For you to get a better understanding of

Bayes theorem, you must recognize that we'll be dealing with sequential events here. This is a case where additional information is acquired for subsequent event. Then the new acquired information is used to analyze the likelihood of the first event.

Consider the sample space S (all the possible outcome in an empirical world). Let ε be a set of all actual observations in an investigation A. let us also assume that no error or inaccuracy will occur, then $\varepsilon \subset S$, this is to mean ε is a subset of S or the point of ε are all found in S.

With this knowledge in mind, now let B denote the hypothesis of the empirical world. If B is true, then all the actual observations will be denoted by ҕ. Since in general B is of limited validity, then all the incorrect observations of hypothesis B, will not be in S.

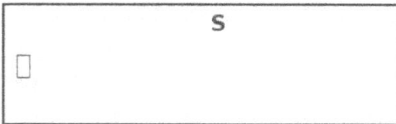

S

ʒ

Illustration 1

Suppose one observation result is in $\mathcal{E}$, that is, it's part of the actual result of our investigation A. What is the likelihood that the result we are probing is in ʒ? Let us see this probability.

From the illustration above we can clearly see that, only $\frac{1}{4}$ area of $\mathcal{E}$, lies in ʒ. Therefore, we can conclude conditional probability that an observation lies in ʒ, given its in $\mathcal{E}$. This can be denoted as $P(B|A) = \frac{1}{4}$, this is obtained as a result of division of $\mathcal{E} \cap$ ʒ (the area of the intersections of $\mathcal{E}$ and ʒ) by the whole area of $\mathcal{E}$. The areas of these regions are directly proportional to their respective probability.

Therefore,

$$P(B|A) = \frac{P(\Box \ \Box \ \Box)}{P(\Box)}$$

....
....
..... EQN1

If we treat probability in an axiomatic way we find that EQN 1 is the formula for conditional probability.

Therefore, we can see that, if we multiply both sides by P ($\mathcal{E}$) we get,

$$P(B|A) \ P(\varepsilon) = P(\varepsilon \cap \mathfrak{z}) \dots\dots\dots EQN2$$

Since EQN2 is an identity, let us interchange ε and $\mathfrak{z}$, we will obtain

$$P(A|B) \ P(\mathfrak{z}) = P(\mathfrak{z} \cap \varepsilon) \dots\dots\dots EQN3$$

The sets $\mathfrak{z} \cap \varepsilon$ and $\varepsilon \cap \mathfrak{z}$ are identical so it easy to note that the left-hand side of both EQN2 and EQN3 are equal, this means that

$$P(B|A) \ P(\varepsilon) = P(A|B) \ P(\mathfrak{z}) \dots\dots\dots EQN4$$

EQN4 is an identity that plays a major role in updating probabilities in Bayesian methods, we will now turn our attention to aspect of updating probabilities. In this case you will find prior and posterior probabilities being used and I hope you still remember what they are.

Suppose a physician estimates the initial/prior probability that a sick father has malaria to be 0.1. We will let the prior probability of this disease to P0 (A), therefore,
$P0 (A) = 0.1 = \frac{1}{10}$

M/N will not simply generalize this prior probability, M represents the number of patients known to be ailing from malaria, in a large population N. additional factors may lead to the doctor judging the disease to be of higher, for example the patient may have had recently toured Benin, or the patients fever may be worst.

To clarify the extent of the whole issue a blood test of the patient is done, and to confirm the fears it turns out to be true.

However, on the basis of recorded past experience, its known that this blood test is not infallible. To support this evidence, it has been recorded that 9 out of 10 people with malaria turns positive to the blood test. Moreover 2 out of 10 people without this disease show positive results to the blood test.

Therefore, we have that,
$P(B|A) = 0.9$ and $P(B|\bar{A}) = 0.2$

Before we go to the computations it's good to note that,
$P(B|A) + P(B|\bar{A}) \neq 1$

Indeed, no reason can be defined to describe as to why the sum of these two probabilities should be equal to one or unity. We now introduce the aspect of the posterior probability.

$P(A|B)$ is defined as the updated/posterior probability, which is the probability of the patient to have malaria. Given that a positive outcome of the blood test was recorded, the conditional probability of having malaria will be show as, and remember this is as per the Bayes law or theorem.

$$P_1(A) \equiv P(A|B) = \frac{P(B|A) \cdot P_0(A)}{P(B|A) \cdot P_0(A) + P(B|\bar{A}) \cdot P_0(\bar{A})} \quad \dots\dots\dots\dots\dots\dots\dots\dots\dots\dots\dots\dots\dots\dots\dots$$

$$\dots\dots\dots\dots \; EQN5$$

We are assuming that the doctor hits some actual level of consistency although we know that,

$$P(B|A) + P(B|\bar{A}) \neq 1$$

Since the prior probability of malaria is estimated by the doctor to be 0.1, this translates to the probability of its absence being 0.9. Then, we generalize this and denote that,

$P_0(A) + P_0(\bar{A}) = 1$, which is very true, therefore

$$P_1(A) = \frac{P(B|A) \cdot P_0(A)}{[P(B|A) - P(B|\bar{A})] P_0(A) + P(B|\bar{A})} \quad \dots\dots\dots\dots\dots\dots\dots\dots\dots\dots\dots\dots\dots$$

$$\dots\dots\dots\dots\dots\dots\dots\dots\dots\dots\dots EQN6$$

If we substitute for the current values, we will end up getting $\frac{1}{3}$, this is after an improvement on our prior estimate by only $\frac{1}{10}$.

Using our previous knowledge on conditional probability, you notice that, the denominator of EQN5, can be substituted with,

$P(B|A) \cdot P_0(A) = P(B \cap A)$ and $P(B|\bar{A}) \cdot P_0(\bar{A}) = p(B \cap \bar{A})$,

From this we can actually write the denominator of EQN5 to be

$$P(B \cap A) + p(B \cap \bar{A}) = P([B \cap A] \cup [B \cap \bar{A}]) = P(B \cap [A \cup \bar{A}]) = P(B) \dots\dots\dots\dots\dots\dots\dots\dots\dots\dots\dots\text{EQN7}$$

Here we are able to add and equate probability because of the kolmogrov axiom of probabilities (the last of these axioms clearly states the rules on additivity of probabilities in case of exclusive events union). And this has been the case for the denominator of our EQN5.

Hence we will simply write EQN5 as,

$$P(A|B) = \frac{P_0(A).\ F(B|A)}{P(B)}$$

$$\dots\dots\dots\dots\dots\dots\dots\dots\dots\dots\dots\dots\dots\dots\dots\dots\dots\dots\dots\dots\dots\dots\dots\dots\dots\dots\dots EQN8$$

This equation flows easily from our EQN4, and we can now proudly say we have managed to derive the Bayes theorem. I hope this has been easy for you and you have understood every step that we took. In case you didn't going through this chapter will cause you no harm.

EQN formally added nothing to our Bayes formula, but remember it has played a key role in helping us arrive at EQN8. It actually enabled us to make the correct estimates of P (B). P (B) is the value of probability that the blood test of the patient gave the correct outcome. Let us consider the denominator of EQN6

$$P(B) = [P(B|A) - P(B|\bar{A})].\ P0(A) + P(B|\bar{A})$$

$$\dots\dots\dots\dots\dots\dots\dots\dots\dots\dots\dots\dots\dots\dots\dots\dots\dots\dots\dots\dots\dots\dots\dots$$

$$\dots\dots\dots\dots EQN9$$

In our case for this malaria patient we can actually calculate P (B), simply because we know the value of P0 (A) - prior probability, and conditional probabilities P (B| Ā) and P (B|A), hence,

P   (B)   =   [0.9-0.2].   0.1   +   0.2

....

....EQN10

     = 0.27

Example 4.1

In a country, out of the total population of adults, 49% are female. In a survey for credit card usage, one adult is selected randomly.

a) What is the prior probability of selecting a male?
b) Additional information is recorded that the adult selected smoke cigar.

Also, statistics show that 9.5% of male in that country smoke cigar, while 1.7% of the women in this country smoke cigar. Using this additional information, calculate the probability the selected adult being male.

Solution

a) Let, M = MALE, F = NOT MALE (FEMALE), C= CIGAR SMOKER, N= NOT SMOKER

Therefore, before we obtained the additional information in part b of the question, we already knew that, the number of males in the country equals 51% of the whole population. Hence

34

$P(M) = 0.51$

b) Incorporating the additional information we have,

$P(M) = 0.51$ probability the selected person is male

$P(F) = 0.49$ probability that the selected subject is female

$P(C|M) = 0.095$ the probability that selected subject is a cigar smoker and is male

$P(C|F) = 0.017$ the probability the selected person smokes cigar and is female

Let us now apply our EQN8 (Bayes formula)

$$P(A|B) = \frac{P_0(A). \; P(B|A)}{P(B)}$$
$$= \frac{0.51. \; 0.095}{P(B)} \; ,$$
$$= \frac{0.04845}{P(B)}$$

Remember $P(B) = [P(B|A). \; P0(A) + P(B|\bar{A}). \; P0(\bar{A})$, hence

$= [(0.51. \; 0.095) + (0.017. \; 0.49)]$

$= 0.05678$

From this calculation we get that,

$P(A|B) = \frac{0.04845}{0.05678}$

$P(A|B) = 0.8532934132$

Before the posterior probabilities (additional information), it was already recorded that 51% of the population in this country is males. After the additional information this probability changed to 0.853.

This probability can be interpreted to mean that 0.853 of the selected survey subjects were cigar-smoking male. There is an increase in this probability, which is very true since the additional information proved that more male than women smoke cigar. Therefore, we do expect the probability to rise rather than decrease.

Exercise 4.1

The Australian health department reports that the rate of HIV virus for "at-risk" population is 10%. Hold some factors constant, when the HIV preliminary test is done 95% Of the time it yields positive results. What is the probability that the randomly selected patient is infected with HIV virus if prior show they tested positive to the screening?

Simple Probability

Probability is defined as the measure of the unpredictability of an event, it's the way we express the likelihood of any event occurring. Let's break this down, if an experiment is performed over and over (using a similar procedure), we can easily record the number of times event A occurs and the number of times the experiments was carried out. In this case we denote the probability of A as

$$P(A) = \frac{N_A}{N}$$

Where

     ☐   N is number of times this experiment was carried out.

     ☐ N_A is number of times A occurred

     ☐   P (A) is the probability of event A occuring.

This equation simply means, as we continue repeating the experiment, the likelihood of A occuring becomes more probable. The value of P(A) is always a figure between 0 and 1, 0 means A will not occur while 1 mean A is certain.

In case of a certain event Ω, then P (Ω) is equals to 1, and N Ω is equals to N, while for impossible event ϕ, then N ϕ = P (ϕ) = 0. From this knowledge we can conclude that, if any disjoint events A and B, then

N (A U B) = N (A) PLUS N (B)

Using the definition of probability, we arrive at,

P (A U B) = P (A) PLUS P (B)

Since not all events are disjoint, then the following expression is true

P (A U B) = P (A) PLUS P (B) - P (A ∩ B)

Proof

If A and B are two joint events, then

A U B = A U (B\A)

So,

P (A U B) = P (A U (B\A))

Using the definition of disjoint tests, and the knowledge that event A and B\A are two disjoint sets, we have

$P(A \cup B) = P(A) + P(B\backslash A)$

Remember,
$P(B\backslash A) = P(B) - p(B \cap A)$

Therefore
$P(A \cup B) = P(A) + P(B) - p(B \cap A)$

Hope you understood the whole derivation, if you did let us now tackle conditional probability

Example2.1

What is the probability of obtaining an even number when a six-sided dice is rolled?

Solution

The sample space for this experiment will be,
$S = \{1, 2, 3, 4, 5, 6\}$

Then the event, E (even numbers that occurred), will be
$E = \{2, 4, 6\}$

Now we use the classical simple probability formula
$$P(E) = \frac{N_E}{N}$$

So,
$P(E) = \frac{3}{6}$

Hence,
$P(E) = 0.5$

Binomial Distribution – The Basics

The Binomial Distribution can be used any time there are one or more discrete events that have exactly two outcomes. An example of a discrete event is a coin flip, or a roll of a die, or a vote cast by a single voter. Each event happens and is done. This contrasts to a continuous event such as learning a new skill, or the tide rolling in, where there is no clear start or stop.

The binomial equation also only applies to events with two mutually exclusive outcomes. Yes/No, True/False, Success/Failure. Events with 3 or more outcomes would fall under the multinomial equation, which is covered at the end of this book.

The purpose of using the binomial equation is to determine how likely a given outcome is after a series of events. For instance, if you flip a coin 10 times, how likely are you to get 7 heads? If you have a 52% winning percentage at blackjack, how likely are you to be ahead after 1000 hands?

Results from the binomial equation have a characteristic shape, similar to the normal curve, show in the example below

Binomial Distribution - 10 Events - 50% Likelihood

Number Of Times

Number Of Successes

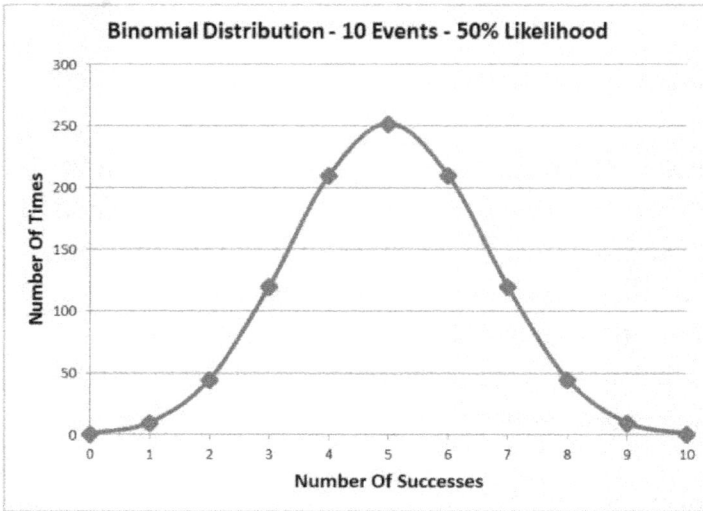

That shape can change as you change the number of events, or the likelihood of a given event. A low probability for an event will cause the bulk of the curve to slide to the left. A high probability for a given event will cause the bulk of the curve to slide to the right.

The first example below starts simple, with only a few discrete events and a 50% likelihood for those events. Subsequent examples add additional complexity step by step.

Example 1 – Binomial Theorem with Pascal's Triangle

In this example suppose that you have a fair coin, and you flip it 10 times. What is the probability that you will get exactly 4 heads?

There are a couple of different ways we could solve this problem, but I'm going to use Pascal's triangle, since it is simple and intuitive. Later examples will use the Binomial equation, since it is more powerful, but harder to remember.

What is Pascal's Triangle?

Pascal's triangle is one way to visualize the outcomes of successive events that can have one of two mutually exclusive outcomes.

For instance, if you have a series of events that were all True/False, or Heads/Tails, or Yes/No and wanted to visualize the total probability of getting a certain number of outcome A or outcome B as you kept doing more and more of the events, a good way to do that visualization is to use Pascal's triangle.

We are showing it here because a series of two mutually exclusive events is exactly what the binomial theorem represents, and hence Pascal's triangle is a good way to show the outcomes from the binomial theorem.

Pascal's triangle starts with the number 1 at the very top row of the triangle. Each row below that has an additional digit in it compared to the row above it. Those digits represent the numbers of A/B outcomes (or success/failure) as you add more events. Here are the first three rows of Pascal's triangle

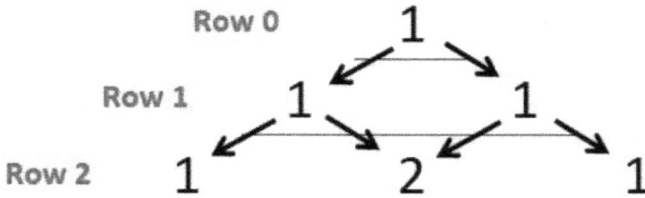

The top row represents having no events, and has only the single digit 1. This is saying that if you don't do any events then there is only one possible outcome (having no successes). The next row, which we have labeled Row 1, has two outcomes. It represents a single event, for instance a single flip of a coin.

The two outcomes are the number of times you will get zero heads after that flip of the coin, and the number of times you will get 1 head after that flip. The next row, which we have labeled Row 2, has 3 possible outcomes. For the coin example, this would be getting zero heads, getting one head, or getting two heads in two flips of the coin.

Note that the total sum of Row 2 is 4. This means that if you flip a fair coin twice, and repeat that pair of flips 4 times, the most likely result is to get zero heads on one pair of flips, one head on two pairs of flips, and two heads on one pair of flips.

However, maybe you don't want to do 4 pairs of flips, maybe you only want to do one pair of flips and want to know your odds of getting a pair of heads, a head and a tail, or a pair of tails. In that case you can normalize the results. The total sum of Row 2 is 4.

If we divide all the values in Row 2 by 4 they become .25, .50, and .25. This means that if you do a single pair of flips, your chances of getting zero heads is 25%, your chances of getting a head and a tail is 50%, and your chances of getting two heads is 25%.

Naturally, Pascal's triangle doesn't stop at Row 2. It continues down indefinitely for as long as you wish to continue calculating it, with each subsequent row representing adding another event. This is the same as saying that you could continue flipping coins and tabulating the outcomes for as long as you wish. You could always do another flip or add another row.

You can calculate each row from the row above it. Each number in a row is the sum of the two numbers directly above it. For instance, the 2 in Row 2 is the sum of the two 1's above it. The two 1's on the edges of row 2 only have a single number directly above them. Those outside edges of the triangle will remain 1 the whole way down.

Here is Pascal's triangle continued down to 5 events

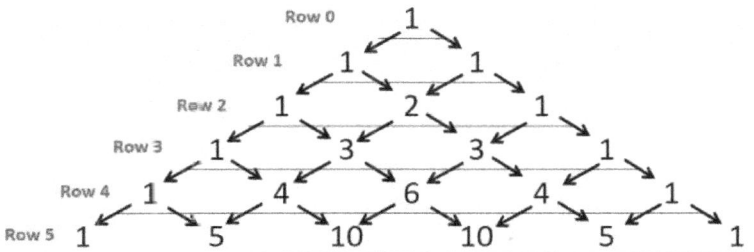

Seeing it extended down this far makes it obvious how the numbers are the sum of the numbers above them, and also obvious that the biggest numbers in the triangle are going to be at the center of any given row, with the smaller numbers on the outside of the row.

Other Ways to Think of And to Visualize Pascal's Triangle

There are a couple of ways to think about Pascal's triangle. The first way is what was described above, each value is the sum of the numbers above it.

The second way is that each number represents the value of the combination formula where a number's row and column corresponds to the number of events and number of successes. This way of utilizing Pascal's triangle is quite useful, and we will go over it later in the book.

The third way of thinking about Pascal's triangle we will touch on briefly here, because it is a very intuitive way of thinking about it.

And that way is that each value in Pascal's triangle represents the number of paths you can take to reach it. For instance, the 2 in Row 2 can be reached by going left and then right, or right and then left. Either of the 3's in the third row can be reached 3 ways.

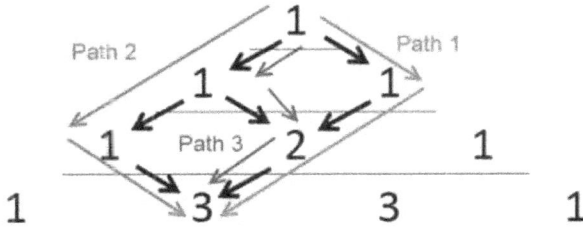

If you think of these paths as events, such as flips of a coin, this is showing that if you flip a coin three times there are 3 different ways that you can get a single head. Each rightward arrow represents a success (heads) and each leftward arrow represents a failure (tails). Each distinct path is a different permutation of the events.

- Path 1 is showing heads-tails-tails
- Path 2 is showing tails-tails-heads
- Path 3 is showing tails-heads-tails

Naturally, for any given number, there are two ways to get to that number, either from the number above and to the left with a rightward arrow, or the number above and to the right with a leftward arrow. (Note this isn't true for the ones on the edges, there is only one path to get to them). So, the number of paths you can take to reach a number is the sum of the number of paths you can take to reach its two parents.

Reshaping Pascal's Triangle for Easier Charts

The visualization of Pascal's triangle shown above, where each row is offset from the row above it, is probably the most common way to show it. However there have been a number of other visualizations developed as well. One of those is shown below

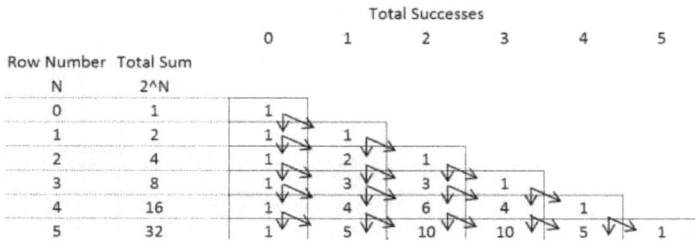

Row Number N	Total Sum 2^N	Total Successes					
		0	1	2	3	4	5
0	1	1					
1	2	1	1				
2	4	1	2	1			
3	8	1	3	3	1		
4	16	1	4	6	4	1	
5	32	1	5	10	10	5	1

The values are the same as in the other version of Pascal's triangle. For instance, row 5 has values of 1, 5, 10, 10, 5, and 1. But in this version each number is the sum of the number above it and the number to the left and above it. It happens that this modified format for Pascal's triangle is easier to create in tables such as the ones that are in excel.

For Even Odds – Highest Values in The Middle

We observed that the highest values in Pascal's triangle were in the middle of any given row. Let's look at why that is, using coin flips as an example. If we go down to 3 flips, we see that the values in the row are 1, 3, 3, 1

Since the sum of those numbers is 8, this means that there are 8 possible orders that all the flips could be in. The values in Pascal's triangle and the binomial equation correspond to the Combination values, where order is not important. This is as opposed to Permutation values where the order is important. If we list out all 8 possible permutations for 3 flips, they are

46

- T T T (0 Heads)
- H T T (1 Head)
- T H T (1 Head)
- T T H (1 Head)
- T H H (2 Heads)
- H T H (2 Heads)
- H H T (2 Heads)
- H H H (3 Heads)

And we can see the same values in these permutations as we did in Pascal's triangle. For either no heads, or all heads, there is only 1 possible way to get that outcome: keep flipping either all heads or all tails. However, if you can have some of flips be heads and some be tails then there are multiple ways to get that outcome, and it occurs more frequently.

The probability of getting either 0, 1, 2, or 3 heads out of 3 flips is the total number of times each outcome occurred, divided by the 8 possible outcomes. This is

- 0 Heads: 1 time, 12.5%
- 1 Head: 3 times, 37.5%
- 2 Heads: 3 times, 37.5%
- 3 Heads: 1 time, 12.5%

With 10 events, and a 50% likelihood for a single event, we can pull the distribution of binomial results from row 10 of Pascal's triangle

Row Number N	Total Sum 2^N	Total Successes										
		0	1	2	3	4	5	6	7	8	9	10
0	1	1										
1	2	1	1									
2	4	1	2	1								
3	8	1	3	3	1							
4	16	1	4	6	4	1						
5	32	1	5	10	10	5	1					
6	64	1	6	15	20	15	6	1				
7	128	1	7	21	35	35	21	7	1			
8	256	1	8	28	56	70	56	28	8	1		
9	512	1	9	36	84	126	126	84	36	9	1	
→ 10	1024	1	10	45	120	210	252	210	120	45	10	1

Those results are plotted below

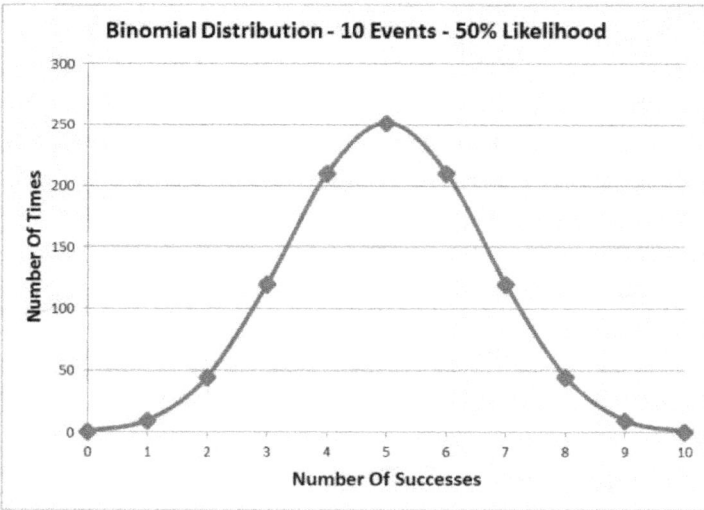

With a 50% likelihood, the most frequent results are in the center of number of outcomes. After 10 flips, there are 1024 possible outcomes, and 252 of those outcomes have 5 heads / 5 tails. So, if you were to guess the number of heads after 10 flips, you would guess 5 as the most likely outcome, even though it is only a 24.6% likelihood overall.

As the number of events increases, the overall shape remains more or less the same, however it spreads out farther to the right to encompass more possible outcomes. Since there are more possible outcomes, the likelihood of any individual outcome decreases.

The chart below shows the binomial distribution at a 50% likelihood for 6 events, 10 events, and 14 events. The distribution for each line has been normalized so that the total probability (area under each curve) is 1.0. You can see how adding more events stretches out the distribution and makes it less tall.

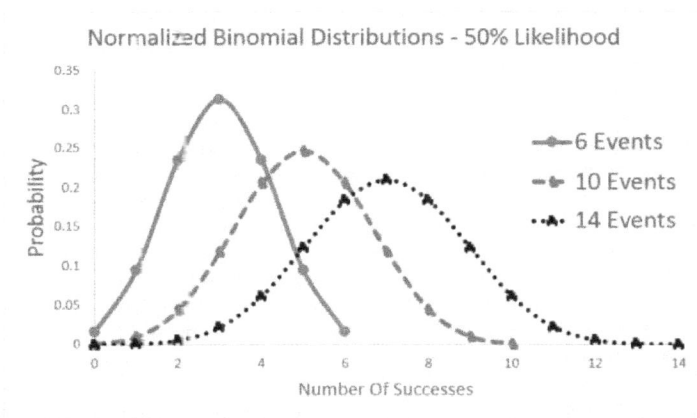

One interesting observation is that this binomial distribution looks a lot like a normal distribution. This is a fact that we will take advantage of later in the book in order to approximate binomial results without having to calculate all of them for large numbers of events.

Non-Equal Likelihood

We've highlighted several times that the shape of the chart changes for different probabilities of success. For instance, the lower the probability of an individual success, the lower the number of successes there will be. In the chart below we have included the binomial distribution for 10 events where each individual event has a 30% likelihood

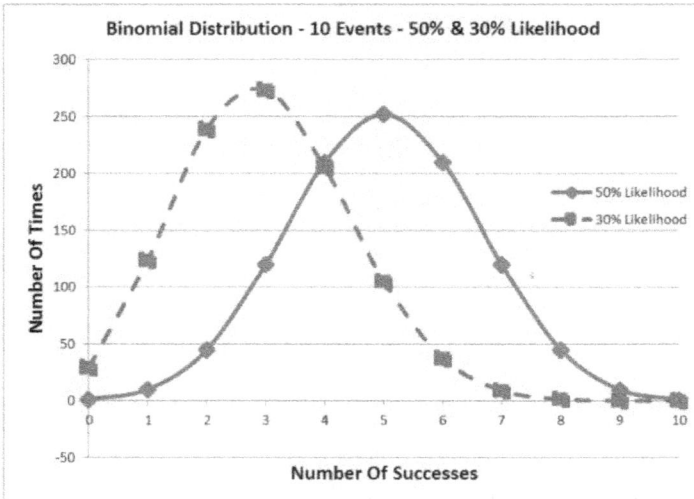

As you can see, this shifts the distribution of outcomes to the left. Importantly however, the full range of outcomes doesn't change.

Even with a 30% likelihood, you can still have 10 successes in 10 trials. No matter how small or large the likelihood of an individual event, as long as it isn't 0% or 100%, the binomial distribution of outcomes will go all the way from 0 successes to the full number of possible successes.

Of course, as the number of events increases, the odds of getting 0 successes or all successes becomes vanishingly small.

This Version of Pascal's Triangle Only Works for Equally Likely Outcomes

It is important to note that the Pascal's triangle we were using above is only valid for events that have equally likely outcomes. I.e. ones where the probability of success or failure is 50%. A coin flip is a good example of this, which is why it was used as an example several times. If the outcomes are not equally likely then you cannot use this version of Pascal's triangle to directly calculate binomial results, although variants are still useful.

The next section looks at a whimsical "Real Life" application of Pascal's triangle, and the following section looks at a variant of Pascal's triangle for uneven odds, before showing how to avoid Pascal's triangle entirely and just use the Binomial equation.

A "Real Life" Example of The Binomial Distribution in Action

Plinko

One game that uses the binomial distribution is "Plinko" in the television game show "The Price Is Right" In this game contestants drop a puck into a big board that has a number of rows of pegs. At each peg, the puck bounces left or right until it lands in one of several slots at the bottom labeled with different amounts of money. At the time of this writing, the amounts of money that can be won in the slots is $100, 500, 1000, 0, 10000, 0, 1000, 500, 100.

A left or right option at each peg replicates what we see in Pascal's triangle. The distribution of results for Plinko pucks can be predicted from the binomial distribution.

We know that the most common slot to land in (if the puck starts dead center) is the $10,000 slot in the middle. However, the two $0 slots on either side are also quite common. Still, from the contestant's point of view, the winning strategy is to start with the puck exactly in the middle, which gives them the largest likelihood of getting the $10,000 prize.

Bean Machine

The Bean Machine was a late 1800's invention that used this same bouncing off a peg property to generate a random number in the binomial distribution. One example of a bean machine is shown here

Photo by Antoine Taveneaux CC BY-SA 3.0

This device used the same principle of left-right choices to divide the balls into slots resulting in a binomial distribution.

Binomial Theorem with Uneven Odds

As promised, it is time to look at events that do not have an equal chance of success. To simulate this, instead of flipping a coin we will think of rolling a die.

A typical die has 6 sides. The odds of rolling a given number on that die are 1 in 6, so 16.6%. To make the numbers a little bit more round, let's assume that you are using a 4 sided die, which do exist. A success for you is to roll a 4, anything else is a failure.   That means the success rate is 1 in 4, so 25%

Why is a rolling a 4 considered a success? Well truthfully, mathematicians, such as you and I, don't concern ourselves overly much with "How can we relate this to the real world?" kind of thoughts, but for these series of problems we can assume that we are simulating a role playing game, such as Dungeons and Dragons, and you need to get hits with your 4 sided die.

You roll the 4-sided die 7 times, what are the chances of getting exactly 2 hits?

We can solve that question directly with the binomial equation, which is

$$f(k, n, p) = \binom{n}{k} p^k * (1 - p)^{n-k}$$

function of k, n & p

Probability of a single event

Number of successful events

Number of unsuccessful events

Shorthand for combination formula a.k.a. results from row n & column k of Pascal's triangle

Probability of a failure

Where n is the total number of events, k is the number of successful events, and p is the likelihood of a successful event on a single trial. (Note, n over k inside parentheses at the front of the problem is mathematical short hand for the combination formula)

However, before diving into that equation, it is once again worth looking at the problem in terms of Pascal's triangle. Previously, we set up Pascal's triangle such that each number was the sum of the two numbers directly above it. Including the number above and to the left in the sum simulated having a success on the most recent event. I.e. if you go from 2 events, 1 success to 3 events, 2 successes you have had a success on the most recent event. Including the number directly above (or above and to the right) in the sum simulated having a failure on the most recent event. For example, if you go from 2 events, 2 successes to 3 events, 2 successes you have had a failure on the most recent event.

Because the probability of success and failure were the same in the previous iteration of Pascal's triangle, we didn't apply any weighting to the chance of success vs failure. Here is a version of Pascal's triangle where each subsequent event assumes a 75% chance of failure, and a 25% chance of success on the latest event.

| | Probability of Success | 0.25 | | | | | | |
| | Probability of Failure | 0.75 | | | | | | |

| | | | | Total Successes | | | | |
Row Number N	0	1	2	3	4	5	6	7
0	1.000							
1	0.750	0.250						
2	0.563	0.375	0.063					
3	0.422	0.422	0.141	0.016				
4	0.316	0.422	0.211	0.047	0.004			
5	0.237	0.396	0.264	0.088	0.015	0.001		
6	0.178	0.356	0.297	0.132	0.033	0.004	0.000	
7	0.133	0.311	0.311	0.173	0.058	0.012	0.001	0.000

To answer our question, we can see from row 7 column 3 of this triangle that the odds of getting exactly 2 hits out of 7 rolls with the 4-sided die is 31.1%

This version of Pascal's triangle was simple to generate. Each cell is 75% of the value above it added to 25% of the value above it to the left. An immediate difference that you will notice in this version of Pascal's triangle vs the standard version at a 50% probability is that these numbers are smaller than one, and they are also decimals instead of integers.

The fact that these numbers are decimals instead of integers is simply because an arbitrary fraction does not necessarily have a clean integer ratio, so it is often simpler to work with decimals. Pascal's triangle with even odds of 50% could easily scale as 1+1=2.

That is why in the normal version of Pascal's triangle each row is double the value of the row above it. In this modified triangle, the fact that the numbers are smaller than 1 is because the sum of each row is equal to the sum of the row above it, i.e. 1.0. This is different than the standard Pascal's triangle.

In the standard version, the sum of each row is double to row above it, but then you divide by 2 to the power of the row number to normalize it. If we wanted, we could multiply each row by 2 to the appropriate power to get a scaled value.

This version of Pascal's triangle was intended mainly to highlight how each subsequent cell is formed. Namely that every directly downward step is equivalent to multiplying by .75, and every downward and rightward step is equivalent to multiplying by .25. In addition to those multiplications, they are also multiplied by the standard values in Pascal's triangle which correspond to how many routes could be taken to get to a given cell.

That means that to calculate the value of any cell in this triangle, we simple need to know what row it is in, and how many successes we have had.

The value of any cell in this modified Pascal's triangle is equal to

- The value of that same cell in the standard Pascal's triangle. Multiplied by
- The probability of success raised to the power of the number of successes. Multiplied by
- The probability of failure raised to the power of the number of failures

If n is the number of events, and k is the number of successes, the formula for any cell in this modified Pascal's triangle is

$$f(k, n, p) = \binom{n}{k} p^k * (1 - p)^{n-k}$$

An intuitive meaning of this equation isn't obvious at first glance, however it turns out to make quite good sense. The next section breaks down and explains the equation so it can be more easily understood.

Exercise 2.1

1.   What is the probability of obtaining two heads, when a coin is tossed? 0.25

2.   When two dices, are rolled find the probability that,

   a) The sum is than 13 (1)

   b) The sum is equal to 4 (1/12)

Intuitive Bayes Theorem

The above solution demonstrates that Bayes theorem can be used to calculate key figures using its formula. However, it very unfortunate that this calculation using the Bayes formula leaves plenty of room for mistakes and errors, a factor that contributes to the complication its use.

Luckily, we have another way of calculation that is less error-prone, easier and more intuitive.

Assume convenient values for all the items/subjects involved. Then construct a table of columns and rows and assign each cell an individual value/frequency based on the given probabilities.

Let's assume a population of 100,000 in the give country, then the total number of male in this country will be 51% of the whole population which is 51,000. The number of male who smoke cigar is then calculated to be 9.5% of 51,000 and is equal to 4845. Therefore, it's clear that out the male adult population 46155 male do not smoke cigar.

The number of female in the country is 49% of the population of adult, and is calculated to be 49,000. Then the number of females who smoke cigar, is 1.7% of 49,000 and is equal to 833. Therefore, out of the total population of adult female 49,000-833 which is equal to 48,167 do not smoke cigar.

All this information can be represented in the table below.

	CIGAR SMOKER (C)	NOT CIGAR SMOKER (N)	TOTAL
MALE (M)	4845	46155	51000
FEMALE (F)	833	48167	49000
TOTAL	5678	94322	100000

Illustration 2

The arithmetic involved in the calculation of this table's data is simple enough to all for a higher degree of accuracy. We partitioned the assumed population, and created a cell for each percentage as represented by the given data.

Now we find the solution to our previous question using the data on our table. We find the probability of the selected subject being a male who smokes cigar. All the conditional probability prescribed in the question above will be put to use here. Our focus will not be on the whole table but rather the cigar smokers' column, therefore we will restrict ourselves to find a male in this column.

The total number of cigar smokers as per the table is 5678, out of this number we have 4845 as the number of male cigar smokers. Therefore, the probability we are looking for P (M|C) will be,

$$P (M|C) = \frac{4845}{5678}$$

Which will give us 0.8532934132. This is the same value as we got using the Bayes formula, and all we did here is introduce an approximate figure of the population. This is more accurate way to calculate the probability of interest and eliminates some of the assumptions made when using the Bayes formula. This way of calculation is called the intuitive Bayes theorem

Generalization of Bayes Theorem

All through this book our discussions have been based on two categories of an event. For example, male and female, smokers and non-smokers, etc. in this chapter we will introduce another aspect of events which include more than two categories. But before we jump into an example to demonstrate this, let us generally take EQN5 into consideration, i.e. we take arbitrary values for $P (B| \bar{A})$, P0 (A) and P (B|A).

This is a case where simple algebra will be used to show some key properties of the posterior probability, $P1 (A) \equiv P (A|B)$. From this we deduce that,

- If $P (B|A) = 1$, and $P (B| \bar{A}) = 0$, then its obvious P (A|B) = 1 if and only of P0 (A) is not equal to zero. This means that, if any test procedure is to be deemed perfect, (which can only work when a disease is present, and doesn't work when the disease is absent), the posterior probability of that disease, when the test records positive outcome is 1. This is to mean that the patient is ailing from the disease. We can anticipate for such a case only if the prior probability we assumed is not equal to zero. Such a case is real and acceptable, because if a doctor is sure the patient is ailing from the said disease, it's obvious he/she would not run the test.

- If $P(B|A) > P(B|\bar{A})$, then $P1(A) > P0(A)$ here we assume that $0 < P0(A) < 1$. In simple words, it the test carried out is more likely to yield positive outcome, when the disease of interest is present than when this disease is absent, then the given posterior/updated probability of the disease is bigger than its initial/prior probability. Here all the case of the limits 0 and 1 are excluded. These are cases when the doctor is initially certain the patient has the disease (1) or doesn't have (0).

- If $P(B|A) = P(B|\bar{A})$ then $P0(A) = P1(A)$. This is to mean if the given test is equally more likely to yield positive outcome, no matter the absence or presence of the disease, then both the prior/initial and the updated/posterior probabilities of this disease are identical. In this case there is no need to carry out the test. We don't need to eliminate the extreme case (P0 (A) =0, and P0 (A) =1) here. This is because they featured in the inequality number two.

These are desirable properties when Bayesian posterior probability is put into consideration. However, in the above example these properties seem less desirable. This is because for the test to yield positive outcome would correspond to the occurrence of the disease, however this all this is dependent on the subjective initial/ prior probability assessment see the above EQN 9 & EQN10.

When we are presented with case of more than two events, then the following conditions must be satisfied,

1.   There is no overlapping, therefore all the events must be disjoint.

2.   When these events are combined they must include all the possibilities i.e. they must be exhaustive.

Example 6.1

Three companies A, B and C are known to manufacture bulbs. Company A manufactures 80% of the bulbs, B manufactures 15% of the bulbs and C makes the remaining percentage. The bulbs manufacture by A 4% defectives, B bulbs are known to contain 6% defectives and those of C contain 9% defectives.

1. If a bulb is randomly selected from the total number of the bulbs, what is the probability it was made by company A?

2. If the randomly selected bulb is tested, and is confirmed to be defective. What is the probability the selected bulb was made by company A?

Solution

Let   A = bulbs made by A
      B = bulbs manufactured by B
      C = bulbs made by company C
      E = defective bulbs
      Ē = not defective bulbs

1. The probability that the selected bulb is manufactured by A, is 0.8, this is simply because company A manufactures 80% of the bulbs.

2. Since we have been given additional probability that the selected bulb is known to be defective. We will revise the probability of part (a), so that the additional information features in our final probability. We will find P (A|E) - the probability that the selected bulb is made by company A and it is defective. The following probabilities can be calculated from the given information,

P (A) = 0.8, because 80% of the bulbs is made by A

P (B) = 0.15, because company B manufactures 15% of the bulbs

P (C) = 0.05, because company C manufactures only 5% of the bulbs

P (E|A) = 0.04, because 4% of the bulbs made by company A are defective

P (E|B) = 0.06, because 6% of the bulbs manufactured by B are defective

P (E|C) =0.09, because 9% of the bulbs made by company C are defective.

We will extend the Bayes formula so that all the three events features, therefore

$$P\ (A|E) = \frac{P\ (A).\ P\ (E|A)}{[P\ (A).\ P\ (E|A)]\ +\ [P\ (B).P\ (E|B)]\ +\ [P\ (C).P\ (E|C)]}$$

Hence,

$$P\ (A|E) = \frac{0.8\ .\ 0.04}{[0.8\ .\ 0.04]\ +\ [0.15\ .\ 0.06]\ +\ [0.05\ .\ 0.09]}$$

P (A|E) = 0.703 (when rounded off)

Tips, Tricks and Trivia to Make Learning Easier and More Fun

In case the examples in the previous chapters do not seem intuitive to you at first glance, you can find some tips, tricks and trivia to help you get motivated in this chapter. Some Bayesians can employ the theorem in just a few minutes. That sounds impressive but you should realize that it also took them hours, days or weeks of studying, practicing and solving using the theorem. This should serve as one of your motivations.

Component of Statistics Subjects

Bayes' Theorem is taught in many Statistics subjects. If you already encountered it before but did not understand it, you might want to go back to your books and notes to find additional explanation. You may also look for problems in the said books where you can try applying the theorem. In case you haven't, you can simply ask books or notes from relatives and friends who took or are currently taking Statistics subjects.

Paper and Pen

Some people find it easier to master if they write it down instead of using computers and smartphones all the time. If you are unsure whether you belong to such group or not, you might want to try solving using paper and pen first. There is no harm in doing so.

Online Calculator

There is no shortage of Bayesian calculators online. These are free of charge and their user interface is simple. These are easy to use provided that you are certain about the probability of the actual event plus the two conditions.

If there are online calculators, why do you have to do the lengthy computations by yourself?

If you truly want to master Bayes' Theorem and apply it in real life, learning how to use the theorem without the calculator is the only way to do so. But then again, you can use the online calculators to check whether your answers are right or wrong.

However, you have to be careful when entering values in the respective slots in the calculators. You have to determine the probability of the actual event you want to know. Also, you need to pinpoint whether the problem intends to measure the probabilities for the two positive results or for the two negative results.

Bayesian Conspiracy

If you feel like too lazy to continue learning Bayes' Theorem, the existence of Bayesian Conspiracy might interest and motivate you to press on. The Bayesian Conspiracy is believed to be the premier group of all practitioners of the theorem.

The said group involves different nationalities from various fields. They are believed to be secretly controlling the publication and other aspects in the application of Bayes' Theorem.

Joining the Campus Crusade in your high school or college is one of the best ways to be accepted as member of the Bayesian Conspiracy. There are rumors that there are only 9 individuals who belong to the topmost level of the group. The 9 figures are known collectively as the Bayes Council.

Bayesattva

Of all the Mathematical theorems known to mankind, nothing is as controversial as the Bayes' Theorem.

Aside from the existence of the Bayesian Conspiracy, there are other urban legends such as the existence of Bayesattva. The name sounds like a demigod in Hinduism.

Based on the urban legend, one who is an expert in Bayes' Theoremis like a demigod indeed. He or she has the power to create and enter an alternate universe with only three things: a computer program, equipment and the mastery of Bayes' Theorem.

In spite of such power, some experts in the theorem chose to stay here and help others understand the theorem. The one who decides to stay and help non-Bayesians and beginners is known as Bayesattva.

Steps

Do not forget the 5 steps in using the basic form of Bayes' Formula and 7 steps for the alternative form. You can memorize the steps by associating them with keywords like.

For the basic form:

Step 1: problem
Step 2: facts
Step 3: table
Step 4: solution
Step 5: probability
For the alternative form:
Step 1: problem
Step 2: facts
Step 3: table
Step 4: values
Step 5: solution
Step 6: probability
Step 7: conclusion

Shortcut

Once you master the 7 steps, you can already try the 3-step shortcut. This shortcut is as follows:

Step 1: Know the prior probability.
Step 2: Get probability for true positive and true negative.
Step 3: Compute.

The Joys and The Challenges of Applying Bayesian Reasoning

Bayes' Theorem is probably the most popular theorem out there. Compared to Pythagorean Theorem which is also another popular theorem, Bayes' work is deemed to be more useful in a variety of fields. In fact, a popular show even featured the theorem in one of its episodes.

It is indeed satisfying to learn and apply Bayesian Reasoning. But then again, there are various challenges in applying it in real life.

Challenges

One of the challenges with learning Bayes' Theorem is the possible shortage of data. How are you supposed to compute for the probability of an actual event if there are no available data about the probability for false positive or false negative results? As a beginner, you may need to research on relevant data on your own. You can make guesses but it takes time before you can make accurate or nearly accurate guesses.

There is also the possibility that the tests where the probabilities are derived from are flawed. So, when you choose data to use in case you want to know the probabilities of an actual event, strive to get the data from at least three studies. There are many studies you can find online but make sure the studies you pick are reliable. Also, you should always check the resources that the researchers used in the said studies.

Joys

Once you have mastered the theorem, you can have an idea on how to make proper guesses about the probabilities of the two positive or two negative conditions. You can assign the probability based on what you have experienced before.

Every now and then, you might encounter people who claim that what they believe is right. Once you go Bayesian, you can prove that their beliefs might not be as accurate as they thought so. You can find satisfaction in proving them wrong. But then again, you might want to act like a true Bayes attva: someone who has a mastery of Bayes' Theorem and helps enlighten others on the workings of the theorem.

Fad or Not

It is indeed fun to learn something new and be good at it, especially if it something as dreaded and loved as Bayes' Theorem. However, some people think that the so-called Bayesian Reasoning is merely a fad. They argue that it will soon fade into oblivion like other school lessons that were deemed to be useful in the real world.

But then again, there are several professionals who can attest to the usefulness of the theorem in their respective fields. You can also find a lot of testimonials of Bayesian practitioners who apply the theorem in their day-to-day activities.

It might just be hype when others proclaim that Bayes' theorem is the answer to the most complex problems or mysteries in the universe. However, the theorem will indeed give validation to your skepticism. Another good thing about the theorem is that it makes you more mindful of the probabilities instead of acting impulsively.

Did you receive a positive result in a medical test for a rare disease recently? Are you sure that your smoke detector still works efficiently? The theorem will help you respond to these issues in a much better way.

Conclusion

Thank you again for downloading this book!

I hope this book was able to help you to understand Bayes Theorem.

Always practice the 7 steps as suggested in this book until you can solve without writing or typing it. Keep in mind that when you need to measure a probability that a test result is true, you need to know the probabilities for the actual event and the two conditions. Additionally, make wise decisions based on the probabilities.

That is the real use of Bayes' Theorem anyway. Weigh on the probabilities carefully and then decide on the best response.

Finally, if you enjoyed this book, then I'd like to ask you for a favor, would you be kind enough to leave a review for this book on Amazon? It'd be greatly appreciated!

Thank you and good luck!

Last Chance to Get YOUR Bonus!

FOR A LIMITED TIME ONLY – Get the best-selling book *"5 Steps to Learn Absolutely Anything in as Little As 3 Days!"* by Edward Mize absolutely FREE!

Readers who have read this bonus book as well have seen huge increases in their abilities to learn new things and apply it to their lives – so it is *highly recommended* to get this bonus book.

Once again, as a big thank-you for downloading this book, I'd like to offer it to you *100% FREE for a LIMITED TIME ONLY!*

To download your FREE copy, go to:

TeachingNerds.com/Bonus

Final Words

I would like to thank you for downloading my book and I hope I have been able to help you and educate you on something new.

If you have enjoyed this book and would like to share your positive thoughts, could you please take 30 seconds of your time to go back and give me a review on my Amazon book page!

I greatly appreciate seeing these reviews because it helps me share my hard work!

Again, thank you and I wish you all the best!

Disclaimer

This book and related sites provide information in an informative and educational manner only, with information that is general in nature and that is not specific to you, the reader. The contents of this site are intended to assist you and other readers in your education efforts. Consult an expert regarding the applicability of any information provided in our books and sites to you.

Nothing in this book should be construed as personal advice, legal advice, or expert advice, and must not be used in this manner. The information provided is general in nature. This information does not cover all possible uses, actions, precautions, consequences, etc. such as loss of data or hardware failure.

You should consult with an expert before applying anything in this book. This book should not be used in place of learning from a professional or seeking advice from a technical specialist.

No Warranties: The authors and publishers don't guarantee or warrant the quality, accuracy, completeness, timeliness, appropriateness or suitability of the information in this book, or of any product or services referenced by this book, other books, and websites.

The information in this book and on relevant websites is provided on an "as is" basis and the authors and publishers make no representations or warranties of any kind with respect to this information. This site may contain inaccuracies, typographical errors, or other errors.

Liability Disclaimer: The publishers, authors, and other parties involved in the creation, production, provision of information, or delivery of this book and related websites specifically disclaim any responsibility, and shall not be held liable for any damages, claims, injuries, losses, liabilities, costs, or obligations including any direct, indirect, special, incidental, or consequences damages (collectively known as "Damages") whatsoever and howsoever caused, arising out of, or in connection with the use or misuse of the site and the information contained within it, whether such Damages arise in contract, tort, negligence, equity, statute law, or by way of other legal theory.

www.ingramcontent.com/pod-product-compliance
Lightning Source LLC
Chambersburg PA
CBHW030534210326
41597CB00014B/1139